RED FOXES

MARYSA STORM

BLACK
RABBIT
BOOKS

Bolt Jr. is published by Black Rabbit Books
P.O. Box 227, Mankato, Minnesota, 56002.
www.blackrabbitbooks.com
Copyright © 2021 Black Rabbit Books

Grant Gould, designer; Omay Ayres, photo researcher

Names: Storm, Marysa, author.
Title: Red foxes / by Marysa Storm.
Description: Mankato : Black Rabbit Books, [2021] | Series: Bolt Jr. Awesome animal lives | Includes bibliographical references and index. | Audience: Ages 6-8 | Audience: Grades K-1 | Summary: "With simple text, clear images, and labeled graphics, teaches beginning readers about red foxes and their awesome animal lives"– Provided by publisher.
Identifiers: LCCN 2019048166 (print) | LCCN 2019048167 (ebook) | ISBN 9781623104528 (hardcover) | ISBN 9781644664384 (paperback) | ISBN 9781623104825 (ebook) | Subjects: LCSH: Red fox–Juvenile literature.
Classification: LCC QL737.C22 S7827 2021 (print) | LCC QL737.C22 (ebook) | DDC 599.36/2–dc23
LC record available at https://lccn.loc.gov/2019048166
LC ebook record available at https://lccn.loc.gov/2019048167

Image Credits
Age Fotostock: Tierfotoagentur / m.blue-shadow, 16–17; Alamy: Henry Ausloos, 5; Premium Stock Photography GmbH, 18–19; Dreamstime: Geoffrey Kuchera, 12; iStock: ABBPhoto, 6–7; NatGeo Image Collection: ERLEND HAARBERG, 22–23; SERGIO PITAMITZ, Cover; Shutterstock: archetype, 14–15; basel101658, 20–21; Eric Isselee, 4, 10, 13, 21; Irina Strelnikova, 3, 24; Jim Cumming, 8–9; Mark Caunt, 1; Petri Lopia, 10–11; Rostislav Stach, 20–21; Vladyslav Starozhylov, 7

Contents

CHAPTER 1

A Day in the Life

A red fox runs through the grass. Suddenly, it stops. It sniffs the air. Then the fox spots a mouse. It waits for the perfect moment. Then it **pounces**, catching the mouse. It's time to eat!

pounce: to suddenly jump toward something

red fox ◄ ··· ········ ··· ···

COMPARING
WEIGHTS ···

Quick Hunters

Red foxes are good hunters. Their powerful noses let them smell other animals. Their big ears let the foxes hear them. They then attack with their sharp claws and teeth.

American shorthair cat
up to 15 pounds
(7 kg)

Red Fox

fur

tail

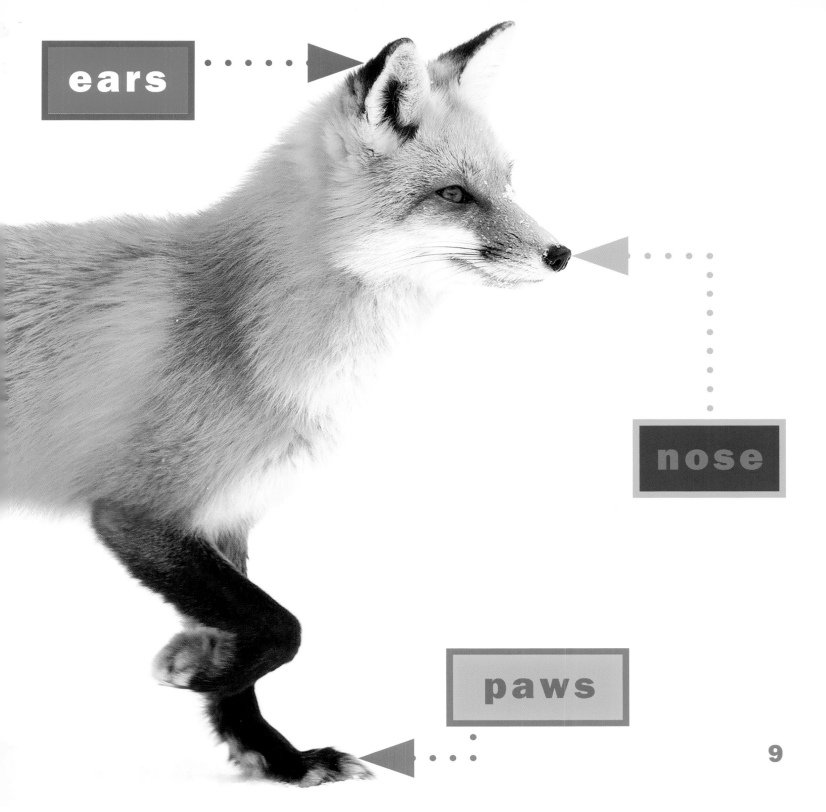

ears

nose

paws

9

CHAPTER 2

Food and Homes

Red foxes mostly eat meat. But they're not picky eaters. They'll eat fruit and seeds too. Red foxes also munch on **insects** and worms.

insect: a small animal that has six legs

FACT

Red foxes sometimes eat garbage.

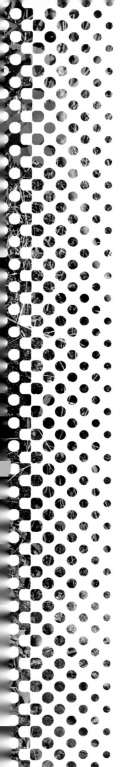

At Home in the World

Red foxes live in many habitats. Some make their homes in fields and forests. Others live in deserts or near farms. They have babies in **dens**.

den: the home of some kinds of wild animals

Where Red Foxes Live

KEY

 = where red foxes live

North America

South America

Europe

Asia

Africa

Australia

CHAPTER 3

Family Life

Red foxes hunt alone. In fact, they spend most of their time alone. Males and females do pair up, though. They start families together.

FACT

Foxes often pair up in the winter.

Growing Up

Female foxes have babies in the spring. At first, the parents care for their pups. The little foxes soon learn to hunt for themselves, though. By fall, most pups leave their parents. They're ready to start their own families.

Newborn Red Fox's Weight

about .2 pound (91 grams)

Bonus Facts

They live two to four years.

They can run up to **30 miles** (48 kilometers) per hour.

Red foxes have 42 teeth.

Few **predators** hunt grown red foxes.

predator: an animal that eats other animals

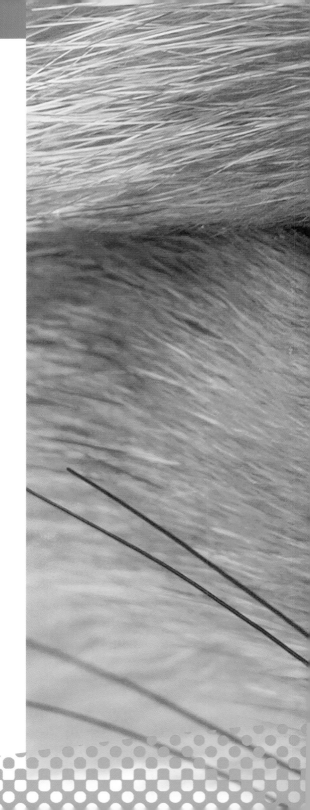

READ MORE/WEBSITES

Jenkins, Martin. *Fox Explores the Night.* Somerville, MA: Candlewick Press, 2018.

Kissock, Heather. *Red Foxes.* Little Backyard Animals. New York: AV2 by Weigl, 2019.

Meister, Cari. *Do You Really Want to Meet a Fox?* Do You Really Want to Meet ... ? Mankato, MN: Amicus, 2019.

Red Fox
www.biokids.umich.edu/critters/ Vulpes_vulpes/

Red Fox
www.nwf.org/Educational-Resources/ Wildlife-Guide/Mammals/Red-Fox

Red Fox Facts
www.dkfindout.com/us/animals-and-nature/dogs/red-fox/

GLOSSARY

den (DEN)—the home of some kinds of wild animals

insect (IN-sekt)—a small animal that has six legs

pounce (POWNS)—to suddenly jump toward something

predator (PRED-uh-tuhr)—an animal that eats other animals

INDEX